ACKNOWLEDGMENTS

Thanks to all who patiently listen to us include "AI" in every sentence and allow us to take conversations into strange and sometimes outrageous new futurescapes.

Thanks to the extremely talented community of geeks, fanatics, and entrepreneurs keeping us all abreast of what's happening in the insanely fast-paced world of AI tools and technology development.

Last, thanks to those who have been so steadfast in encouraging and motivating us as we lean ever further into the frightful and thrilling changes of this fourth industrial revolution.

USE OF ARTIFICIAL INTELLIGENCE

Humans developed this product with the support of AI tools and technology. This included brainstorming topics and outlining, narrative inputs, graphic development, and iterative reviews to improve the final product. While the technology doesn't yet exist to "write me a book" (or not a good one, at least not at the time this book was written), we took every advantage we could to develop a high-quality product… as should you.

ABOUT FRONTIER ACADEMY

Frontier Academy was built by working professionals looking to harness AI technology to stay relevant in their industries. Our mission is to help individuals and teams stay competitive in rapidly evolving workplaces by leveraging the exponential productivity enhancements these tools offer.

Yes, we also offer training and consulting services.
If you're looking for more hands-on support, use the QR code below and give us a good way to contact you.

Learn more at FrontierAcademyLLC.com!

CONTENTS

A NOT-SO-DISTANT FUTURE

Fresh out of college and newly hired as a marketing manager, Sarah starts her second week with a mix of excitement and nerves. She sips her morning coffee while her AI assistant briefs her aloud on the day's agenda, sorts and reviews her messages aloud, and helps her set major goals for the day. She has her first assignment – to launch an important new product.

Her AI assistant, already familiar with the company's best practices and industry standards, presents a detailed, personalized plan for her. More than that, it highlights the similarities between this project and the one that went so well during her senior year of college. It also detects her anxiety and offers reassurance that there are resources to help during every step in the process. It reviews each aspect of the plan with Sarah, explaining each step and patiently answering questions until she feels comfortable. Sarah feels empowered to inject her own creative customer engagement concept, which they assess together. In the end, they decide together to change the plan slightly to incorporate the ideas she offered that are most likely to improve their results.

By lunch, Sarah and her AI have generated a full range of slogans, ad copy, visual concepts, and marketing materials that match her company's branding guidelines and reflect best practices from a wide range of successful campaigns for similar products – including funny taglines sure to drive customer engagement. They review these together, tweaking them to suit Sarah's unique style while still matching the best practices they were designed with.

While Sarah has her lunch, the AI assistant begins the campaign on her behalf, pushing out materials to their selected markets and working within their pre-defined budget for ad space and platform use. It notifies her when social media comments begin coming in that warrant a human touch, and Sarah finishes her lunch to begin responding based on the assistant's recommendations on what might be most appropriate to say.

Customer engagement continues for the next couple of hours, with the AI assistant picking up on Sarah's tone and style and allowing her to "one-click" respond to hundreds of comments in near real-time. All the while, the AI assistant uses company-proprietary metrics and systems to track engagement levels, customer sentiment and feedback, and product sales information, offering tweaks to Sarah's responses to help her keep trends in the right direction.

By mid-afternoon, the campaign is well underway, and Sarah is able to reduce her involvement to a minimum. Her AI continues to respond to comments and engage customers in the background, giving Sarah a real-time look into the results of her launch project.

Her AI assistant reminds her of an upcoming meeting with her manager and offers her a pre-developed report that showcases the plan she implemented, the unique tweaks she made, the style used for customer engagement, and the resulting engagement levels and sales numbers. It highlights a new best practice to consider implementing on other similar campaigns. They review the report together and make a few updates based on what Sarah knows her manager will want to see. Sarah approves the report, and it's sent to her manager in time for her afternoon meeting.

~~~

Sarah's story illustrates a very near-future workplace where AI acts as an advanced personal assistant, anticipating needs, providing expert guidance, and helping navigate complex tasks with ease.

Sarah's productivity is now limited by her comfort with the tools she was given, and these tools exaggerate the unique value she can add to each task.

This isn't a distant, far-fetched scenario; it's a realistic glimpse into how generative AI is already beginning to transform how we work. The practical applications of AI are enabling employees like Sarah to achieve tangible results quickly and make informed decisions confidently. This is the future of modern business, and it's quickly approaching.

1: HANG ON TIGHT

When stormy seas around you rage,
Steady your soul, turn the page.
- ChatGPT, 2024

Before you dive into Chapter 2 and start rolling out a comprehensive AI program at your company, we strongly recommend at least a quick glance through this chapter. It absolutely will not make you an overnight expert in the technology, but we hope it at least gives you a big-picture view of what's been happening... and why so much fuss is being made over it.

First, it's important to understand that this chapter was written on the same shifting sands as the entire book. Leaps in AI-related computing power continue to be made and continue to be astounding, to the point where planning begins to feel like a fruitless exercise. In fact, this entire book is being written with that in mind in an attempt to embrace this change with a mindset of embracing change in general. For you, this means you will very likely need to adjust strategies and tactics almost as fast as you can make them.

Second, there are a host of books, podcasts, YouTube channels, and other sources that will help you go much deeper than here. This is the executive summary, written only to give context. We'll offer some links and recommendations for other sources throughout the book to let you dig further if and when you feel so inclined.

The Elephant In The Room

Why does it suddenly feel like AI is everywhere? Haven't we had versions of this for decades?

Much of the answer is because of what happened on November 30, 2022, when the elephant officially walked into the room, and OpenAI released ChatGPT 3.5 to the public.[1] It was a quiet release, a beta test, likely meant to get some outside opinions on the utility of their new software program. However, by January of 2023, OpenAI had more than 100 million active users, setting a record for the fastest-growing platform in history.[2]

The rest of the answer is more complex and related to a curve of technological advancement so increasingly steep as to give anyone looking a strong sense of vertigo. Using the NVIDIA V100 (2017, 125 TFLOPS) and their GB200 NVL72 platform (2024, more than 1 EFLOP) as benchmarks, graphics processors are a staggering 8,000 times faster than they were five years ago, reducing the time needed to train large AI models by similar orders of magnitude.[3] The seeds of natural language processing and machine learning, neither of which are new technologies, have now given rise to multi-modal generative tools on the verge of rivaling human intelligence. The AI of 6 months ago is not the AI of today, and every industry is feeling the effects.

Consider the emotional impact. Remember how powerful ChatGPT felt the first time you tried it (if you haven't, now is the time to read our other book, *Embrace Your Existential Crisis*). It was like jumping forward in time from typewriters straight to word processors, experiencing the backspace and spell check and font styles all at once. And you can imagine how long a company relying on typewriters for office functions would last when competing with companies that use word processors. More than that, you can now

---

[1] https://openai.com/index/chatgpt/

[2] https://www.reuters.com/technology/chatgpt-sets-record-fastest-growing-user-base-analyst-note-2023-02-01/

[3] https://developer.nvidia.com/blog/nvidia-gb200-nvl72-delivers-trillion-parameter-llm-training-and-real-time-inference/

imagine how everything that can be done with a computer, which is almost everything, will be able to be automated. It's shocking to your primal "there's a tiger in the bushes" core, inducing both your fight and flight instincts at once.

Consider also the broader financial and security aspects. According to hostinger.com, the global AI market is expected to reach $306 billion by the end of 2024 and grow by 37% or more every year until at least 2030. Unsurprisingly, these figures have been accompanied by a flurry of investing, with funding for AI globally reaching $20 billion in February 2024.[4] A company called Pika (a new AI video generation startup company) made the news with their $80M in funding being raised within only 14 months of being founded.[5] According to tracxn.com (a startup company tracker), there were more than 75,000 AI-related companies worldwide in May 2024.[6]

On a related note, one important reason AI is in the news so frequently is that the content creators who drive most of what we see online were some of the earliest and most aggressive adopters of AI tools. The world has been shocked to find out how well AI tools can generate highly creative content and images. And marketing teams have been the first and biggest beneficiaries, with their productivity multiplying and entire job categories almost vanishing overnight. It isn't easy to go anywhere on social media or online without running into it.

This business and marketing explosion has been tempered only by the recurring appearance of AI in the news related to job loss fears, data protection concerns, and the use of AI tools for more nefarious purposes (using deepfake image and video creation, voice emulation, and misinformation propaganda). The capabilities of these new tools started a wave of anxiety, much of which may be

---

[4] https://www.hostinger.com/tutorials/ai-statistics

[5] https://www.washingtonpost.com/technology/2024/06/04/pika-funding-openai-sora-google-video/

[6] https://tracxn.com/d/sectors/artificial-intelligence/__cbMnXfS2GfFo4Vi2dxZyUy7l4O8WyzVYLseb9keW5cI/companies

warranted, that news outlets have been quick to capitalize on.

In response to these fears, our online platforms and governing bodies have also taken plenty of stances and even some steps. Multiple US Government agencies have established AI leadership to help guide the use and protect against the misuse of AI tools in their organizations. In fact, an Executive Order on the "Safe, Secure, and Trustworthy Development and Use of Artificial Intelligence" was released in October of 2023. According to shrm.org, as of May of 2024, California lawmakers are discussing more than 30 AI-related bills aimed at regulating the use, misuse, and potential threats of AI in their state.

Luckily, if you're reading this book, you're likely someone choosing to embrace the changes that are already on our doorstep. We hope you're reacting with a survivalist's sense of urgency rather than falling prey to a paralyzing fear of the unknown. The two feelings are similar, but one leads to stagnation and the other to action.

If so, this means good news for you and your company. The natural language interfaces of these tools and their highly intuitive nature mean that anyone can start taking advantage of them today. In addition, the tools being developed on the back of these technologies are targeted and transformative for a host of use cases. The net result is that the productivity and skill level of employees can jump by orders of magnitude, drastically cutting learning curves for junior team members and magnifying the impacts of expert employees. And these benefits are being made available at costs so low as to make ROI calculations a joke.

To take advantage of these technologies, we believe future market leaders will undergo fundamental shifts in how they think about work. We believe companies will soon be letting go of things like standard operating procedures, talent development programs, and even organizational structures. We believe speed, adaptability, and experimentation will increasingly become the new drivers of success. And in this book, we invite you to follow us down the rabbit hole of this thought process.

Speak Intelligently About AI

As a leader in your company, we get that you're going to have to sprechen sie AI, even if it's just enough to not say something dumb. To that end, we want to walk you through some AI concepts, terms, and misperceptions to at least put a couple of tools in your toolbox.

First, what do we even mean by AI? Are we talking about robots or supercomputers or what?

In this book, we are almost exclusively using AI to refer to multi-modal, Large Language Models (like OpenAI's ChatGPT or Anthropic's Claude) that generate content using a natural language interface. Multi-modal simply means it can generate not only text but images, audio, video, or other types of content. The term "Large language model" (LLM) originated because of the size of the data used to train the system, but it's quickly becoming a misnomer since these models extend so far beyond language. Natural language is exactly what it sounds like – the user interface is designed to emulate conversing with another human. If you've heard terms like Natural Language Processing (NLP) or Machine Learning, these refer to broad varieties of AI tools that have been maturing for decades and have been critical stepping stones in the creation of today's LLMs.

Next, let's talk about their size, because it's easy to imagine how LLMs are really big and really expensive. And it's true that their creation takes millions of dollars, a lot of time, and powerful computing systems to create. But the version you are using online is surprisingly small. Small enough to run on a decent laptop without taking up too much memory. This is part of the magic that these AI companies have discovered… the probabilistic connections between concepts can be stored in a small space and still offer tremendous value.

A term you've likely heard often is "training," and this is the process of building a large AI model by feeding it as much information as possible so that it can identify as many connections as possible. Since training is so costly and not (currently) a continual process, this is why asking AI about current events leads to confusion and misinformation. It only understands the information it was trained on – no more and no less. This is also why "context window" is so important to users, because the concepts often don't include information you need the model to know.

We're talking pretty vaguely about concepts, connections, and probabilities, and that might be worth explaining. Some like to compare it to your phone's built-in "next word predictor" feature that uses probabilities to guess what you might want your next word to be when typing a message. But let's do a quick visualization to help further.

Imagine watching a romance movie about a rich man and a poor woman who meet unexpectedly on a train – both played by high-profile actors. You have already guessed the plot, and that's because you've seen enough movies (been trained on a large amount of data) to have a reasonably accurate guess at how those types of movies should go (understanding of probabilistic connections). You haven't seen that exact movie before, but you could almost write the rest of the script. The couple will fall in love, face some big obstacles that stand in their way, and eventually work things out nicely. In this scenario, your brain is the AI tool, wired over many years to make connections between similar scenarios and be able to predict with reasonable accuracy what the audience expects to see next.

For you and your company, this means that you are better off thinking of AI as a synthetic human brain trained on large amounts of public information. And like the human brain, it has limitations that highly tailored software solutions do not. For example, current models have a hard time with information retrieval and even telling the difference between facts and fiction (challenges we all face as humans). However, like a human brain, it excels at telling stories and making creative, even intuitive leaps in understanding between seemingly disconnected ideas.

So what does it mean when they say AI is "hallucinating"? Like a human, it may be great at guessing movie endings but terrible at recalling every line in a single script. We like to say that the models are sometimes "more helpful than accurate" and want so badly to give you an answer that they will say anything that sounds close enough. These responses are not necessarily grounded in reality and are referred to as hallucinations. If you demand that it recite the script for a specific movie, it will give you a script. But it is very unlikely to be word-for-word the same as the original.

What Could Go Wrong?

Not unexpectedly, most mainstream news sources and more than a few great science fiction movies focus on the hazards and dangers of AI, so we felt it important to unpack a few of these. Partly to enable you to speak on these topics when others inevitably bring them up but mostly to take away these excuses for not moving forward. Because even if AI is truly an existential threat to humanity as we know it, you can either spend your time worrying or spend your time preparing. We believe the latter is always more productive.

First, let's get the existential threat of a robot apocalypse off the table. With AI tools so close to emulating humans, they are dipping precariously into the proverbial "uncanny valley" – a term used to describe our strong sense of unease when a human-like object starts to closely (but not quite) resemble a human. Terms like Artificial General Intelligence (AGI) and Artificial Super Intelligence (ASI) are used to describe systems that surpass our understanding and begin pursuing their own objectives. AI enthusiasts are calculating the probability of doom (p-doom), where AI takes control of our lives and decides we aren't worthy of planet Earth. Or they hope for a utopia where AI has solved all our problems once and for all. These are fascinating discussions that ultimately you (1) can't do anything about and (2) offer no serious insight into how best to run your business in the meantime. Aside from moving your operations into a bunker on another planet or starting to sell bunkers here on Earth, this threat analysis is pretty unhelpful from a corporate perspective.

Instead, let's talk about job loss, because that's the one with the highest likelihood of hitting home for the largest number of people. In late 2023, an MIT study concluded that ChatGPT increased productivity in highly skilled workers by 17% and up to 43% for lower-skilled individuals.[7] And anecdotal stories suggest targeted tools are having a multiplying affect in some professions. Frontier Academy's other book, *Embrace Your Existential Crisis,* includes a thorough analysis of this risk for individuals, and you might consider sharing some copies with your team. For companies, there are two risks here: (1) the risk of losing market share to your competitors and needing fewer employees as you shrink, and (2) the risk of products not being affordable if job losses create broad economic challenges. For the former, getting the team up to speed is doable and necessary. For the latter, you can use automation to decrease costs to accompany a decline in your market's purchasing power. In other words, addressing these risks requires companies to lean further into AI and automation, not less.

Next are the risks common to any major software system, including data protection, ethical use, and regulatory compliance. Addressing this requires a robust framework for vendor assessment and rules that protect information and can respond with speed without creating unreasonable barriers to progress. A very delicate balance, to be sure. At a minimum, we recommend (1) paying the few extra dollars for a protected subscription, (2) implementing a standard practice of reviewing terms and conditions for tools (e.g., *Will your data be used to train future models? Will you have a secure, proprietary workspace?*), and (3) providing simple training or communication on restricted uses (e.g., *Do not use for unethical purposes, such as replicating the likeness of a known person. Do not use to store personal information.*).

Last but not least, we think there's a real risk of self-inflicted wounds resulting from over-engineering solutions based on today's AI tools. It's a risk because it's so incredibly tempting. New companies are being formed that leverage today's technology and then vanish overnight with some future release of ChatGPT. If you spend your company's time and resources year re-engineering all of your processes based on today's technology, you may have to start all over again before you finish the first round. We are experiencing

---

[7] https://mitsloan.mit.edu/ideas-made-to-matter/how-generative-ai-can-boost-highly-skilled-workers-productivity

changes faster than ever, so the old playbooks need to just be thrown away. In their place will need to be a culture and system of flexibility and adaptability ready to integrate and use new technologies as quickly as they appear.

If it's not clear already, we think the biggest AI risk your company is likely to face (that you can do something about) is falling behind. This can happen by not empowering employees to use AI tools, not reshaping your organization and processes for the future, and not fostering a culture able to handle the coming changes.

Practical Applications

With that background out of the way, the rest of this book offers strategies and tactics to help you initiate and drive forward the changes needed to successfully leverage AI tools and technology – today and in the foreseeable future. In particular, we are focused on the use of the most pervasive Large Language Models like ChatGPT and Claude, also referred to as Generative AI or multi-modal LLMs (we use these terms and "AI" interchangeably throughout the book). We help you find the people, processes, and structure that will foster a culture of AI experimentation and integration.

To set your expectations, we should also say what is *not* included. We are AI enthusiasts, not developers, and so our focus is on the impacts and applications of the technology, not its functionality. We also recognize that a host of AI technologies exist beyond LLMs, including things like predictive analytics, recommendation systems, and adversarial networks, but these all have targeted use cases and primarily affect a handful of very large companies. Our target audience is medium to small businesses where AI tools will have the greatest effect (for better or worse).

But the most important purpose of this book is to give you permission. Permission to drag your organization through the hype and the fear. Permission to look seriously at how AI can help your company grow and thrive. Permission to get started.

2: OPEN THE DOOR

*"90% of the value comes from giving people access
to the tool and not thinking too much about it."*
- Brad Lightcap, OpenAI COO, 2024

That first chapter was a lot, and intentionally so. If you want to make it through this change, a strong sense of urgency will absolutely help you on that path. If you haven't already started, it's time to start moving in the direction of getting yourself and your company prepared. Luckily, those first steps can be, and likely should be, kept small and manageable.

Our recommendation for the first step is to create an environment conducive to freedom of exploration. Let people discover the power on their own, and let them find their own way to extract value from it. The beauty of these AI tools is that they conform to the individual user's need, so there's no longer a need to drive everyone toward some sort of unifying best practice!

According to a 2024 Microsoft and LinkedIn survey on AI use at work, 75% of knowledge workers are using AI at work, and roughly 80% of those are using their own tools. [8] And these AI users report time savings, increased focus on key priorities, better creativity, and

---

[8] https://assets-c4akfrf5b4d3f4b7.z01.azurefd.net/assets/
2024/05/2024_Work_Trend_Index_Annual_Report_663d45200a4ad.pdf

more enjoyable work. But more than half of these "are reluctant to admit to using it for their most important tasks."

In other words, professionals are *already* seeing the value and using it to improve their performance, they just aren't telling you about it. To take better advantage of this phenomenon, we recommend simply giving them the access and trust needed to begin openly and collaboratively experimenting. By giving them access intentionally, you instantly (1) create a more open environment for discussing the best ways to integrate tools into company processes and (2) enable AI users to be engaged in protecting critical data and mitigating other AI tool use risks.

That's Chapter 2 in a nutshell, so we won't blame you if you jump to Chapter 3. If you decide to keep reading, the sections below include a host of benchmarked best practices to open the door even further for employees while reducing risk to your company.

What Others Are Doing

One interesting way to discover how AI is being incorporated into business operations is to watch job requisitions. A quick search will reveal thousands of recent job postings that have "artificial intelligence" included in them. Some describe AI-based tools they are expected to use, some have it included in the description of the company's strategic direction, and some are looking for people to help navigate this new landscape with them.

At Frontier Academy, we fully expect to see AI tools increasingly become an inherent part of future job requirements for knowledge workers in the way that Microsoft Office or the Adobe Creative Cloud have been in the past. Not only is AI simply being integrated into most existing software platforms, but the communication skills required to effectively interface with the tools will be increasingly critical to their performance.

In many cases, companies have elected to create new AI positions entirely to help understand and meet this future need. IBM does a great job on its website describing a Chief Artificial Intelligence Officer, or CAIO, as an executive "focused on overseeing the development, strategy, and implementation of AI technologies."[9] Even federal agencies in the US are jumping onto this trend, creating their own CAIOs to take advantage of the technologies while also ensuring they can be used responsibly.[10]

So, do you need to create a CAIO position? This depends on the size of your organization, the level of change needed, and a host of other factors. Once Generation Alpha (born after 2010) starts entering the workforce, the discussion will likely be moot since they would have grown up with it. Until then, to get through this transition period, you'll absolutely need advocates on your executive team and early adopters being groomed throughout your organization. But more on that later.

Most companies are starting to use AI tools in organizations or functions where it has the biggest and most obvious impacts. Teams that have heavy communication requirements, such as Human Resources, are finding these tools to be life savers. In fact, a Harris Poll conducted for Grammarly reported that "knowledge workers, particularly those in IT, HR, and technology, have overwhelmingly embraced gen AI. They note it has transformed the way they communicate at work (71%) and improved how effective that communication is (68%), and they ultimately believe AI will enhance their work, not replace them (66%)."[11]

As a specific example, many HR teams are using AI to help assess candidates for job openings (a great use of LLMs). An IT company called EXL has this caveat on their job postings to make it clear what they're doing: "EXL may use artificial intelligence to create insights on how your candidate information matches the requirements of the job for which you applied."[12]

We recommend talking to some of your team leads. Ask what's

---

[9] https://www.ibm.com/think/topics/chief-ai-officer

[10] https://www.nasa.gov/news-release/nasa-names-first-chief-artificial-intelligence-officer/

[11] The Harris Poll, conducted for Grammarly and reported in their *2024 State of Business Communication*

[12] https://www.exlservice.com/careers

keeping them from staying focused on their biggest challenges. If it's communication related, they might be a great place to start.

Tools to Start With

Okay, so you found some teams to start with and hopefully even found a few folks enthusiastic about diving in further at work. What now? What tools do they need to use? Are the tools safe to use with your proprietary data?

To be fair, tool selection is a daunting prospect. According to the TAAFT (There's an AI For That) website, an average of more than 550 new AI tools are being released each month, touching roughly 4,800 different job functions, with many jobs having thousands of AIs tailored to related functions (Communications Manager, 5,900+ tools; Data Entry, 4,900+ tools; and the list goes on).[13] Most of these tools are brand new, putting them at higher risk of not being available later, not having strong data protection mechanisms in place, and just not being as reliable.

We've grouped tool options into three rough categories to simplify what could easily become an entire book into itself: AI Frontliners, Tailored Technologies, and Proprietary Power.

The AI Frontliners are the most popular and well-known tools on the market. They are available at a very low cost, have built-in data protection mechanisms and/or options, and are very likely to continue sticking around. The two biggest examples include:

♦ *OpenAI's ChatGPT[14]* – While there is currently a free version of their latest model available, a "Team" subscription for your company gives your folks access while preventing your data from being used to train future models. The subscription version also enables your team to create and share custom "GPTs" internally, making it extremely useful for automating repeat activities at a team level.

---

[13] https://theresanaiforthat.com/

[14] https://openai.com/chatgpt/team/

- *Anthropic's Claude 3 Opus[15]* – Anthropic has several models, with Claude 3 Opus being the most powerful (and arguably more "smart" than ChatGPT in many functions). All of Anthropic's models protect user data extremely well.[16] Some argue that Anthropic will soon become the leader in the AI capability race and that their models are the most safe and secure. Their "Projects" modules allow similar functionality to GPTs for repeat, complex tasks.

- *Google's Gemini[17]* – Gemini is free to try on your Google Workspace (Gmail, Docs, Sheets, etc.), but the paid versions are still relatively inexpensive ($6 to $30 per month) and include bonuses like meeting data storage, business emails, meeting recordings, and custom Gems for repeat work. All are secure, and even the free version offers a simple setting to prevent your data from training future versions.

If you didn't catch it, there are relevant links included as footnotes after each of the tools above. And, of course, a lot more detail on these tools and how to start making the most of them is included in Frontier Academy's *Embrace Your Existential Crisis*.

As far as price goes, the tools above will set your company back some whopping tens of dollars a month per employee. And given the data on productivity and quality improvements, this is likely the lowest cost enhancement your company will ever experience. In fact, just to pile onto the many examples we've given already, the following is an excerpt from Harvard Business School's 2023 working paper on the *Effects of AI on Knowledge Worker Productivity and Quality*: "… consultants using AI were significantly more productive (they completed 12.2% more tasks on average, and completed tasks 25.1% more quickly), and produced significantly higher quality results (more than 40% higher quality compared to a control

---

[15] https://www.anthropic.com/news/claude-3-family

[16] https://support.anthropic.com/en/articles/7996885-how-do-you-use-personal-data-in-model-training

[17] https://gemini.google.com/

group)."[18] And these results are nearly instant, not requiring years of continuous improvement cycles to achieve.

Next, let's talk about Tailored Technologies. These are custom-built tools designed for specific functions.

Some of these tools use their own proprietary AI models, as is the case for Midjourney, which has its own LLM trained on imagery to enable text-to-image content creation. Another big example is Meta AI, which answers questions for Facebook and Instagram users and has some similar multi-modal capabilities to ChatGPT. And the availability of open source LLMs keeps the barrier to entry relatively low for new AI startup companies.

However, the vast majority of Tailored Technologies are products that run the models of others within their software. In particular, many use models built by OpenAI and Anthropic to drive the results desired by their users. The biggest example is Microsoft CoPilot, which is essentially ChatGPT tailored for use within MS Office products.[19] Users essentially have an AI chatbot available to them in their MS Office products to streamline their work directly. Microsoft is also offering Copilot PCs that include ChatGPT embedded in their operating system for a faster and more seamless interface.[20]

The value of these tools is that they (1) make the interface more in line with your existing processes, (2) save you the effort and learning curve for building your own, and (3) increase the consistency and predictability of outputs. For example, Sudowrite helps users walk through a novel development process step-by-step, with AI engagement helping you in each phase. It even lets you select which model to use, depending on your needs.

Tailored Technologies have an extremely wide range of pricing, from free or very cheap (Midjourney starts at $10/month and is

---

[18] https://www.hbs.edu/ris/Publication%20Files/24-013_d9b45b68-9c74-42d6-a1c6-c72fb70c7282.pdf
[19] https://www.microsoft.com/en-us/microsoft-365/business/copilot-for-microsoft-365
[20] https://blogs.microsoft.com/blog/2024/05/20/introducing-copilot-pcs/

arguably the most capable text-to-image model available) to tens of thousands of dollars for more niche products. Because of the prolific and continual release of these niche and exciting products, we recommend treading carefully here in terms of how you spend your time and money. Smaller, tailored AI models may not be kept up to date, making the products less powerful over time compared to their competitors. Tools built on other models (like Sudowrite) add value today, but future versions of ChatGPT and Claude may easily perform the same functions.

You may also be tempted to create your own local, company-proprietary AI model (Proprietary Power). There are services available to do this for you, taking a tool like GPT Enterprise or an open source LLM and training it on your internal data.[21] This is a more costly endeavor (on the order of $250K and higher) but has the potential value to do much more tailored work. PwC famously spent upwards of $1B over three years on rolling out GPT Enterprise across their 75,000-person operation, including an "AI Factory" to provide training and generate enthusiasm.[22] As a reseller for ChatGPT Enterprise, PwC naturally advocates a much more aggressive deployment playbook, including the significant adjustments in data management and cybersecurity needed to get here.[23]

Since you're reading this book, you're likely part of a small or medium-sized business and very unsure where to start. Our recommendations are to (1) get some folks on your team access immediately and (2) give them access to the AI Frontliners like ChatGPT and Claude, for now. Once your team discovers the power and potential impacts of AI tools for your specific company challenges and needs, the next logical step will be spending additional time and money on tool research.

---

[21] https://openai.com/index/introducing-chatgpt-enterprise/

[22] https://www.pwc.com/us/en/tech-effect/ai-analytics/generative-ai-impact-on-business.html

[23] https://www.pwc.com/us/en/tech-effect/ai-analytics/guide-to-generative-ai-for-the-cio.html

Encouragement, and Plenty of Rope

If you haven't already guessed, we believe the first and most important step is to open the door to AI for your company or team. The Grammarly report we cited previously also identified that "Gen AI experimentation appears to drive new interest, with 58% of knowledge workers wishing their organizations were more open to AI implementation, especially those in the IT (80%) and technology (76%) sectors, where the benefits could be substantial."[24] The same study notes that Meaning, if you want to get things going, start by making it clear that it's not only acceptable but encouraged.

So, what does encouragement look like? We'll leave you with three simple, common-sense strategies to consider.

First, your executives and the direct supervisors of any potential users need to be on board. Ensure executives at the highest level understand basic AI terms and technology so that this "terra incognita" (unexplored territory) feels a bit less daunting to them. Clearly articulate the potential improvements to communication, graphics development, customer service, and more using this book, one of the myriads of studies cited within it, or a more targeted study you find online. In other words, get them through their fears and leaning toward enthusiasm. If you want to take it further, you might even consider setting a percent quota for leadership in getting their personnel to individually request access (with no rationale needed). Whatever you do, make sure the top brass understands the stakes and can help start the ship moving in the right direction.

Second, ensure anyone interested is given access to the tool or tools you've decided to use. As we noted above, the cost of the AI Frontliner models is absurdly low. Give a company credit card to someone and empower them to sign up anyone that asks. Anytime access is granted, give the employee some clear instructions to set aside time to plan, learn, and experiment. Ensure their supervisors

---

[24] The Harris Poll, conducted for Grammarly and reported in their *2024 State of Business Communication*

are aware that their personnel are helping research the future of the company in real time and to help them create a safe space to do so. This "permission to explore" will be at the heart of the organizations able to adopt these tools most quickly and effectively.

Third, offer some early and very basic internal support. Assign someone to give a quick presentation that debunks myths about AI and offers recommendations on how to keep company data secure for the selected tool(s). Prepare to answer the inevitable question, "Will there be training for this?" with a list of helpful YouTube videos and platform tutorials. Assign mentors and protégés within teams to encourage early adopters to train more nervous team members. Encourage team brainstorming challenges (using AI) to identify challenges AI could help solve and ways to get started.

If all else fails (no, we do not apologize for the shameless plug here), give all your folks a copy of *Embrace Your Existential Crisis* and call us for an initial consult to see if we can help. The book was written for the explicit reason of helping individuals interested in AI to start learning and experimenting. And Frontier Academy was formed to help both individuals and companies work through this change.

All three strategies are really aimed at a single overriding objective: to build an adaptive, AI-intelligent organization that can roll with the changes that will continue to come. By creating a safe environment for agile experimentation and individual creativity and by focusing first on letting individuals find small ways to improve their productivity and quality, you'll be heading in a strong direction to achieve even bigger and more pervasive benefits over the long term.

3: GET THE BALL ROLLING

*"The only thing to fear with AI is your competition
figuring out how use it more effectively than you."*
– Jonathan Mast, White Beard Strategies

Now that your folks have access to a couple of generative AI tools, it's time to think about change management. About building up some organizational momentum toward your future-proof team. Maybe there's a distant future in which AI just does whatever you can think for it to do, but for now you'll need people at multiple levels of your organization.

Fortunately, the psychology and science of change management are nothing new. Over the last century and a half, models of change have been tested and refined that address how social systems and organizations are truly interconnected, how individuals contribute to or slow down progress, how to create urgency and shift cultures, and even how neuroscience comes into play (e.g., the amygdala fear response versus the positive dopamine motivator). The only thing that's really different here is the pace of the change and the degree of potential impact. Instead of trying to reinvent that wheel, the next few chapters stand on the shoulders of other experts. We essentially use their models and terms to put our recommendations into context to help you design a change management strategy applicable to your team and organization.

In this chapter, we walk through the models and terms to make sure we're on the same page. We then lay out recommendations for some important first steps, including setting up a change leadership team, inciting some intentional anxiety, and baking in some team resilience to what's coming.

The Frameworks of Giants

This section will give you a quick overview of some tried and true change management frameworks and terms. But it's important to note that this is not a change management book, and we do not recommend starting to plan a robust bureaucratic movement. We're using the philosophies and terms to help organize our recommendations, but in the face of this fast-moving technology, this change game is better called Speed. With speed, the focus shifts from monitoring to movement, from perfection to progress. And with progress will come the learning pains that are the real secret behind future-proofing your team.

In fact, here's a fun alliterative saying you can put on an inspirational poster (to replace the one about Teamwork in your conference room): *Painful progress is preferable to postponed perfection.*

In all seriousness, the philosophies below provide some extremely helpful context for Chapters 3 to 6, so bear with us in quickly defining them. As an aside, we strongly encourage you to continue learning about effective change management from other sources. It can do nothing but help you.

Starting out at 50,000 feet, the Gartner Hype Cycle is one of the most helpful graphical ways to look at the maturation and adoption of technology on a broad scale.[25] If you aren't familiar with them, Gartner, Inc. is a well-known research and advisory firm and has proven the accuracy, validity, and utility of this cycle over the last ~30 years. The Gartner Hype Cycle was introduced in 1995 to better understand and help organizations navigate the rapid evolution of technology in the 1990s. Then and still today, it's extremely useful in

---

[25] https://www.gartner.com/en/research/methodologies/gartner-hype-cycle

expectation setting, strategic planning, and even risk mitigation. The cycle has five phases, each of which are summarized, and poorly over-simplified to maximize their usefulness in this book, below.

1. Innovation Trigger – A product breakthrough or launch generates a lot of interest, but the technology is largely unknown or unproven.

2. Peak of Inflated Expectations – Early adopters begin to experiment, and rumor mills generate hype around some exciting success stories.

3. Trough of Disillusionment – Further use reveals unexpected limitations and setbacks, while only the most committed continue leaning forward.

4. Slope of Enlightenment – Technology advancements get closer to the original hype, and adoption expands as the real value becomes increasingly clear.

5. Plateau of Productivity – Stable technology and widespread acceptance, combined with standard best practices, yield consistent and predictable value.

Next, we drop down to 10,000 feet, where the team resides. At this level, we like psychologist Dr. Bruce Tuckman's self-named "Tuckman Model" to describe the stages of organizational adoption.[26] Dr. Tuckman first introduced these terms as part of a psychological bulletin in 1965, written to connect the dots between previous research and create a more comprehensive model for understanding how small groups evolve in response to change. These stages are extremely useful in helping manage conflicts and resistance to change, guiding communication strategies, improving training effectiveness, and measuring the success of change management initiatives. Each stage is briefly defined below.

1. Forming – A period of orientation and acquaintance when the team is getting to understand each other's roles and their

---

[26] Tuckman, B. W. (1965). *Developmental sequence in small groups.* Psychological Bulletin, 63(6), 384-399.

new objective(s).

2. Storming – Conflict and competition are the hallmarks of this important stage, when tensions rise as individuals question the team's goals and their own roles.

3. Norming – Teams begin to overcome their conflicts by finding new ways to collaborate and adapt.

4. Performing – The team operates efficiently toward its goals, feeling more comfortable with the change, their roles, and a shared vision.

5. Adjourning (aka Mourning) – The goals having been achieved, the new approach has officially become "the way" and the change-oriented team is no longer necessary.

Down at the individual level, one of the most widely used and accepted frameworks comes from the 1962 book *Diffusion of Innovators* by Dr. Everett Rogers.[27] Dr. Rogers was a statistics and sociology professor who had been studying how ideas and technologies spread across different communities. His book is foundational in the study of innovation diffusion and was updated as recently as 2003 to incorporate 40 years of observation and insight. It describes factors that influence the rate of change adoption and uses an S-curve of change to place individuals into the five categories described below.

1. Innovators – Venturesome individuals willing to take risks, try new ideas, and initiate the adoption process.

2. Early Adopters – Respected leaders with the social influence needed to validate and promote the innovation to their colleagues and a broader audience.

3. Early Majority – Caution but forward-leaning individuals willing to adopt new ideas before the average person. They provide the critical mass needed for an innovation or change to become mainstream.

4. Late Majority – Skeptical individuals who need a lot of

---

[27] Rogers, E. M. (2003). *Diffusion of Innovations* (5th ed.). Free Press.

assurance, peer pressure, or even economic necessity to reluctantly participate.

5. Laggards – Traditionalists who are very resistant to change and will continue to use the old methods or tools until there is literally no other choice.

In addition to their convenient "5-part" nature (so you can easily remember them on one hand?), each of these frameworks has something to offer in understanding the complex social and psychological dynamics of change. And there are no doubt studies already being done on how to navigate change in a post-AI world. But our advice, one last time for the people in the back row, is this: If you want to stay ahead of your competition, don't get bogged down in a change management exercise. Keep speed and progress as your top priorities.

Form a Change Leadership Team

To get change moving at your company, we first recommend pulling together a team that can lead this change. Feel free to call them a Tiger Team, Task Force, Visionary Ventures Vanguard, or whatever you like. But be sure they include a combination of Dr. Rogers' Innovators (risk takers) and Early Adopters (influencers). Consider including someone on the team who has a data security background to minimize both risks and perceptions of risks related to protecting company information. This team will also benefit greatly from executive head cover, so consider identifying a champion who can clear any bureaucratic hurdles, provide funding, and generally keep the access door open for them.

When creating the team's charter, make it clear that their goal is NOT to roll out a new technology. They should be building a nurturing a culture where experimentation and failure are rewarded and where new technologies can be safely researched and tested. With technology changing so quickly, the true goal is preparing for the next iteration and the one after that, not wasting time redesigning processes around last year's tool. Include in the responsibility to go after small, short-term goals, as this quick-turn mindset will act as a model for the longer-term cultural change objectives.

If you read the previous section, you've probably already spotted that we're deep in the Forming stage of Tuckman's Model. And if you recall, the stage after this will be a challenging one: Storming.

To prepare for the Storming, be sure to set your change leadership team on the dual paths of learning and teaching. With technology moving so quickly, simply keeping up with the changes is a difficult task by itself, and it will take time and effort to do well. However, the model of being a continuous learner will need to start with this team. They should also prepare to share what they learn with others, find relevant training materials online, or build their own. The further the team gets into this, the more questions they will need to be ready to answer.

When you think about training, we want to caution you to keep the molehills small (not make mountains out of them). Since the interface of these generative AI tools is natural language, the interface is as easy as talking. This means that most training will effectively be "permission to try" sugar pills designed to get people finding their own best ways to use the tools.

And not to be repetitive, but Frontier Academy's *Embrace Your Existential Crisis* does a pretty good job covering early training topics and offering resources to help people get started.

Incite the Innovation Trigger

Speaking of existential crises, the reason we titled our other book that way is because we've found no better way to describe the feeling you get when you first discover the power of generative AI for yourself. A sense of awe and excitement followed quickly by feeling like your livelihood and very humanity may be at stake. Maybe it's because so many AI movies fall into the thriller or horror genres. Or maybe it's because of how natural it feels to "talk" to the computer. Regardless, there's an eye-opening moment that occurs, leaving you with an unsettling sense of anxiety about where the technology is going and how it will affect your life.

And the truth is that it very well could affect the lives of most knowledge workers. According to the SEO.ai report on 2024 employment job statistics, "30% of workers worldwide fear that AI might replace their job within the next three years," and that number is 74% in India.[28] Some lower skilled jobs are already being replaced, and wages for those types of jobs are also lowering. The good news in the same report is that 81% of office workers hold a favorable view of AI, and 51% said that AI enables them to strike a better work-life balance.

Our advice is to pull this Band-Aid right off. Have everyone use their generative AI account for a simple function like explaining how it can help them at their job (example below). This doesn't take any serious time, forces them to engage with the tool, and is very likely to get them both anxious and excited at the same time. This is absolutely natural and exactly the response you're hoping for.

Eye-Opening Prompt Idea

"Ask me a series of questions in an interview-style format to understand my role at my company.

Once we're done, create a table that includes a list of challenges I probably face in my job and corresponding ways generative AI can help resolve those challenges."

Done correctly, you can encourage most of your team to turn their immediate anxiety into a healthy "sense of urgency." We love this phrase because it helps you mentally look at anxiety as a source of potential productivity. You might also consider repeating some of the same things AI enthusiasts have been saying, like "This is a tool to make your job and life easier," "it's like you'll have your own super smart assistant all the time," and "learning to use AI is the best way to avoid being replaced by it." These are all true statements, which is exactly what makes them so valuable.

---

[28] https://seo.ai/blog/ai-replacing-jobs-statistics

In addition to giving them a sense of urgency, we recommend the influencers in your change leadership team provide an inspirational presentation to nudge that urgency in a productive direction. It should highlight both the rapid pace of change and the way you're planning to help everyone keep up with it (support, training, and the freedom to experiment). It should also be clear that there is plenty of room for them to help guide and change the future in their areas of expertise. Encourage them to start learning how to effectively communicate with these tools and to adjusting their mindset toward adaptability and continuous learning – those capabilities and traits that will be the highest need going forward.

We realize that not everyone will be immediately inspired, and this will be the time that some of the Storming starts to happen. Fortunately, your team has been getting smart on the tools, preparing answers to likely questions, and developing some simple training offerings to help your most nervous employees. Take the time to work with the most nervous individuals one-on-one, helping them through the process. This will pay off exponentially, as they are probably the same folks who would otherwise be spreading their fears and concerns to anyone willing to listen.

Build Up Organizational Resilience

Before we leave this chapter, we want to offer a few more tips to smooth out the edges of the Storming phase a bit and starting to prepare for long-term culture change. In particular, we mean building up organizational resilience. That is, your team's ability to keep moving forward even when things get tough.

In their spotlight series on managing risk and resilience, the Harvard Business Review does a nice job explaining that highly resilient organizations have efficient processes, simple rules, and the ability to improvise.[29] Below are some thoughts on how to apply these three tenets to your team when it comes to AI.

1. **Ensure processes are clear and efficient**. This is a bit of a

---

[29] https://hbr.org/2020/11/building-organizational-resilience

no-brainer because, of course, you don't want time being wasted, but consider that many knowledge workers may have overlapping duties, shifting priorities, and personal best practices. When major disruptions happen, it will be essential to really understand what's important, who is ultimately responsible for a function, and what key activities need to happen for a successful outcome.

2. **Establish simple rules that hold up to change**. Your company likely has some existing patterns, where certain types of projects get approved, certain clients get priority, etc. Lean into these patterns and document your internal rules of thumb. When things start moving faster, these rules will help act as risk-mitigating guardrails for your company.

3. **Enable and empower improvisation**. Encourage small teams to tackle challenges as a group, like using AI to automate internal communication or reports. Consider adding "brainstorming challenge resolution with AI" or similar topics into early training materials. Identify problem solvers in your organization, and make sure everyone knows who they are and when to call them.

Not to question HBR, but we also have two bonus recommendations. These are consistent communication and planning for your next generation of personnel.

4. **Communicate clearly and consistently**. Set reasonable expectations to help your team mentally prepare for known challenges. Make sure early training materials are available and able to be expanded as needed. In particular, be sure to establish a good mechanism for sharing success stories and lessons learned, as these will help others when they face similar challenges.

Pro Tip: Anyone can start a video call with themselves and hit "record" to capture a best practice or lesson learned. This is an effective and simple way to start building up your repository of generative AI success stories.

5. **Recruit for the future.** Adjust your recruiting processes to look for candidates who are already resilient and adaptable and who have faced unexpected challenges and thrived. During interviews and onboarding, set expectations for continuous learning and experimentation. In short, make sure future waves of personnel are already better suited and prepared for a culture of resilience.

When the Storming comes, and it will, organizational resilience will help you get through it as smoothly as possible. We realize that the road to organizational resilience is unending, but hopefully, these tips will help you get started.

4: TIME TO RAMP UP

"You don't learn to walk by following rules.
You learn by doing, and by falling over."
- Richard Branson

At this point, you're so enthralled by the fun, interesting presentation of our ideas that you haven't even given your team access to any tools yet, much less formed your AI Tiger Team. That's perfectly fine and not a big deal at all... we can wait.

~~~

Great!

It's a good feeling to have the engine of progress humming along nicely within your team, with innovators testing out products, influencers communicating value, and training helping bring the early majority into the fold. In fact, you're likely already hearing about some early successes and exciting new ideas, and you're thrilled to learn they're not just coming from the innovators. And while the phrase "peak of inflated expectations" is still floating around in the back of your mind, you may feel tempted to close the book here and claim victory.

We're tempted to ask for your five-star Amazon review right now while the iron's hot. Instead, and we realize this means you'll forget about that review, we recommend continuing to read. There's still plenty more track on this roller coaster.

This chapter identifies some of the steps you can take during this exciting time to generate the psychological momentum needed to make it through Chapter 5 (called "Through the Valley"). This is a great time for converting some of the early majority folks into early adopters. For getting long-term commitment from the stakeholders that could otherwise slow things down again. And for leaning into the Forming phase by putting time and resources into some of the low-effort, high-return ideas.

Inflate Expectations

If you've personally used one of these tools, you already know how easy it is to get excited about them. Even if there is still some room for improvement, the technology available today has amazing potential to transform your personal and team's productivity in a host of ways. As more and more people in your organization start coming to this realization, a bubble of expectations will be forming that absolutely cannot last.

But don't pop it, not yet. We recommend doing the opposite and leaning into it. Have your change leadership team broadcast some of the early successes. Give the overly enthusiastic innovator the microphone for a bit. Consider more subtle approaches like giving surprise awards to individuals who set up some automation on their own, recommended looking harder at a new tool, or requested to take some training they found online. You can give these awards privately and let their co-workers discover on their own (rather than have it blasted at them) that the company is serious about this.

The goal here is to take advantage of the natural enthusiasm that will occur around this exciting capability as you move toward the Peak of Inflated Expectations. Give that enthusiasm broad attention so that it reaches as far into your organization as it can. That's not to say you should be exaggerating the capabilities, but it's important to let the positive messages take root before going on a full-fledged the-sky-is-falling campaign. And with everyone having some basic access, this will give them permission to look a little further and see how it might help them.

As folks across the organization start using the tools, they will feel much more vested in making them work.

Initiate Pilot Projects

The next logical question is: how do we deepen that feeling of being vested in generative AI tools? Our recommendation is to inspire a deep curiosity about the extent of their capabilities as it relates to how they can help each individual personally.

Dr. George Loewenstein, a well-known behavioral economics scholar, is often cited for his "information gap" theory – that feeling of deprivation from needing to know something you don't – as a driving force behind nearly all types of curiosity.[30] This theory continues to prove itself every day in marketing campaigns ("You won't believe what's inside!") and offers a glimpse of insight into why tools like ChatGPT have exploded in popularity so quickly. That is, once you start talking to what feels like the smartest person you've ever met (on every topic), the number of questions you have becomes endless.

One way to tap into this curiosity is by implementing some small projects to add value to your organization. Fortunately, brainstorming is one of the superpowers of AI tools, and the process of trying to find the right project will open the curiosity gate pretty wide all by itself.

You might recall our "eye-opening prompt idea" from Chapter 3 to identify challenges that generative AI can help resolve. Here's where that list of ideas can gain ground. Below is a more complex example (that you don't need to use verbatim but can) of a prompt to help someone pick apart one of those challenges and find ways to potentially add value.

---

[30] Loewenstein, G. (1994). The psychology of curiosity: A review and reinterpretation. *Psychological Bulletin*, 116(1), 75-98. https://doi.org/10.1037//0033-2909.116.1.75

Prompt Idea for Solving a Challenge

"Please perform the following tasks in order, waiting for my permission to proceed before continuing.

1. Ask me a series of questions one at a time to understand why [insert a major challenge] is so challenging for me.

2. Once we both agree that the questions have been answered, provide a list of recommended tasks that generative AI can help me perform that could reduce these challenges. Place them in a table with the following columns: Challenge Being Addressed, How Generative AI Can Help, Impact of AI on the Task (High/Med/Low), Complexity of Getting AI Support for the Task (High/Med/Low).

3. Next, I will select a task to work on and offer input on it.

4. Last, once I select a task, draft an outline for a step-by-step plan that guides me through using generative AI to complete the task."

By using prompts like the one above, employees are learning how to think about and communicate the key challenges of their job. And then they're learning about all the ways generative AI might be able to help them. Encourage them to prioritize things that they struggle with the most and those that AI will most likely be able to help. This will help them build some earned confidence in their ability to use the tools.

As another thought, delving into the world of generative AI immediately reveals a host of niche tools ("Tailored Technologies") built on the same technology. Another good example of a pilot project is to investigate one or more of these tools. Have folks listen to a demonstration and share what they learn. Invariably, they are learning not just about the tool but about the capabilities and limitations of the system it's built on. This is another great way to inspire curiosity and deepen that feeling of investment.

Contests and Collaboration

Even with encouraging people to learn about their challenges and review some new tools, it's very likely that your Visionary Ventures Vanguard (see Chapter 3) owns the preponderance of pilot projects in progress. This is natural and for good reasons, but getting through the next phase will need strong engagement from your early majority folks and maybe a few of the late majority.

Yes, we realize breaking this into its own section may be indulgent, but at least it helps break up the monotony.

To get your gears going, below are just a handful of ideas on how to start engaging more folks. Note that we used an AI tool to help with our brainstorming, and so should you.

♦ Minimum Requisite Training – Create a simple training module that requires trainees to engage in simple interactions with a tool and demonstrate personal value. Consider requiring certain skills, positions, and perhaps all new hires to take the training.

♦ Collaborative Workshops – Host facilitated events where solutions to broad company challenges are brainstormed using generative AI. This enables the trainer to draw more reluctant employees into the conversation by monitoring their concerns and how they respond to different activities.

♦ Cross-Department Project – Set complex challenges in front of small teams that span different functions. This not only helps with learning and engagement, it also brings unique perspectives and work styles to figure out how AI can help with real problems.

♦ Role-Swap Initiatives – With generative AI, any employee can work on any challenge. Pair up employees from different teams and backgrounds to identify and solve each other's challenges. You might be surprised at how effective this can be.

♦ Innovation Contests – Depending on what motivates your workforce, contests with meaningful prizes can often draw engagement. Consider either individual or team-based contests conducting pilot projects to streamline cumbersome processes.

Plan for the Long Run

While folks are still Forming and moving through the Peak of Inflated Expectations, we recommend putting some rules and even systems to keep things on track. While most pilot projects won't warrant detailed tracking, you may want to monitor the bigger efforts and at least collect the impacts of others. You will definitely want mechanisms for sharing best practices, lessons learned, and successes as they occur. Something as simple as a page on your corporate SharePoint will work great. You might also update a policy or two to ensure experimenters keep their access and permission to do so.

We won't get into project management recommendations here, but keep in mind the rapidly changing nature of this technology. In fact, projects might be better classified as "experiments" since they will come and go so quickly. We recommend keeping things simple and flexible, or the resources you pour into project tracking might undo much of the return on your investment.

You have probably noticed our silence on bigger organizational and structural changes that might bear fruit. As the technology continues to mature and evolve, those types of changes may be critical to your company's survival, and we dive deeply into this subject in Chapter 7. First, let's focus on making it to the other side of the Trough of Disillusionment.

5: THROUGH THE VALLEY

"Failure is simply the opportunity to begin again,
this time more intelligently."
- Henry A. Ford

In screenplays that follow Blake Snyder's "Save the Cat" construct, there is a pivotal moment before the final act when the protagonist faces impossible odds, and the edge of your seat feels suddenly more comfortable.[31] This moment is ominously called the "Dark Night of the Soul," and it's when their true determination is put to the test. Fortunately, this is exactly what they have been preparing for, and they can move with clarity, confidence, and resolve to turn this adversity into a thrilling opportunity.

The building of excitement and expectations in Chapter 4 was fun, but reality is nearly guaranteed to fall short. The tools are harder to use than they seemed at first. They show promise but can't quite do what you want. They aren't as accurate as you need for business purposes. High traffic levels cause crashes, delays, or reduced capabilities. And a growing number of folks might actively (or passively) push back, intentionally slowing progress in reaction to feeling like they might be replaced.

---

[31] Snyder, B. (2005). *Save the cat! The last book on screenwriting you'll ever need.* Michael Wiese Productions.

All of these and more will happen, and momentum will falter. After all, hype can be tough to live up to, particularly when so much is still unknown about this emerging technology.

But companies are pushing through. In a Gartner survey conducted at the end of 2023, generative AI was found to be the top form of AI solution being deployed in the US, Germany, and the UK.[32] In that same Gartner survey, they identified the primary obstacle to adoption is demonstrating the value of AI projects. They cite issues such as technical difficulties, data-related problems, lack of business alignment, and lack of trust in AI.

This chapter tackles the Storming phase and Trough of Disillusionment head-on, offering suggestions to address common concerns and to continue building a flexible, adaptable, future-proof culture. It also offers advice on how to get the most from the Slope of Enlightenment on the other side of this valley.

Into the Trough of Disillusionment

As the Storming phase begins, and it will, you'll start to notice a drizzle and then a deluge of complaints and concerns, accompanied by a slowing of progress on projects and overall engagement. Team and company leadership, as well as your change leadership team, will be put on the defensive.

When your team first touched generative AI, they were blown away by how intuitive it felt and how fast it gave such thrilling results with only simple commands. But when they need it to do something much more ambitious, it doesn't quite deliver. They spent too many hours trying to tweak and work with the tools, and the result is inching closer but not quite there. At this point, it would have been way faster to do it the old way. What's even the point?

The truth is that their concerns are valid. Generative AI is powerful and exciting, but it has limits like any software system. It

---

[32] https://www.gartner.com/en/newsroom/press-releases/2024-05-07-gartner-survey-finds-generative-ai-is-now-the-most-frequently-deployed-ai-solution-in-organizations

can misunderstand the subtleties in instructions, accidentally perpetuate biases based on its training data, give inaccurate or irrelevant results, or just have a "bad day" due to server loading or other technical issues.

In an excellent, thoroughly researched report put together in 2024 by HFS Research titled *Democratizing GenAI: A reality check for business transformation*, they state that "the most significant challenge lies in disappointment if the expectations of GenAI are overly inflated."[33] They make a strong case for generative AI not being a one-size-fits-all solution and for needing to understand your industry-specific nuances. They also include great advice (on page 15 if you follow the link to read it yourself) from enterprise leaders that align pretty closely with this book.

Instead of letting the late majority and the laggards claim victory here, we recommend patiently addressing concerns and steadily helping your team find the value that is there. Below is a list of some of today's most common frustrations and concerns, along with some simple recommendations on how to address each.

♦ Lack of Internal Support – Some employees might find that their supervisor doesn't support the idea, training wasn't helpful enough, or other priorities just keep them too busy. To address this, consider targeting those supervisors for collaborative workshops, or pull individuals out of the team for temporary assignments related to generative AI testing or role-swap initiatives (Chapter 4). Or ask their peers for recommendations on additional external training they've found to be helpful.

♦ Inconsistent and Insufficient Results – Generative AI tools are *consistently inconsistent*, which can definitely be an issue for folks wanting to automate a very specific activity. In addition, employees might be asking for help with tasks too complex for today's systems. This typically comes down to a misunderstanding of how the tools function, and additional training and experience on how to communicate with them can

---

[33] https://s3.wns.com/S3_5/Reports/Democratizing-Gen-AI-Market-Impact-Study-Report.pdf

help dramatically. Also, be sure to continue to share clear expectations regarding tool limitations and long-term thinking – today's tools can't do what tomorrow's tools will be able to.

♦ Outputs Too Generic – Your top performance and subject matter experts will notice immediately that it can't handle industry-specific nuances or complex tasks except through more effort and iteration than is worth doing. Encourage them to extract what value they can and to keep honing their communication skills while the tools mature.

♦ General Overwhelm – Many employees may stop before they start. There is too much newness and too many options, and they may even feel that their job security is being threatened. These are deep emotions and tough to counter, but consider setting up peer mentorships, recommending additional external training, and offering reminders that the key is to just keep taking small steps forward. It won't hurt to share your strategic objectives and the company's plans to use productivity gains to grow (not reduce staffing levels).

♦ System Reliability – As your team starts relying more heavily on the technology, server overloads or outages might become bigger and bigger issues. This is a good reason to have access to a couple of similar tools/systems. Long-term, this might also be a reason to consider more costly solutions such as Tailored Technologies and Proprietary Power (See Chapter 2).

Centaurs and Cyborgs

That was a lot to think about, so let's take a break for a more fun subject. Namely, understanding a bit about how different people will be using the tools. In 2023, Harvard Business School conducted a study of 758 consultants to assess the impacts of generative AI on their productivity and quality. In doing so, they also identified two different ways of working with the technology and named them Centaurs and Cyborgs. Here's what they said:

"One set of consultants acted as 'Centaurs,' like the mythical half-horse/half-human creature, dividing and delegating their solution-creation activities to the AI or to themselves. Another set acted more like 'Cyborgs,' completely integrating their task flow with the AI and continually interacting with the technology."[34]

In other words, centaurs tend to take more time to set up the tool and enable it to accomplish bigger tasks independently. Examples might be creating a complex prompt to automate a weekly document review or adding search engine optimization to large swaths of web content. Cyborgs, however, tend to use AI like a co-worker, mixing and matching small portions of tasks based on what feels comfortable or who can do it better. Like brainstorming potential Excel formulae to try or helping rewrite an email to fix their tone.

Understanding this helps you support learning in the ways that best work for individual team members. Centaurs may work great on a common AI project with a larger group. Cyborgs may work best solo or in a pair with another cyborg, since both are testing out smaller tips and tricks more quickly and on-the-fly.

On a related note, centaurs, in particular, may start looking to integrate AI into workflows and processes. Given the quick maturation of the technology, we recommend against this temptation. Applaud their enthusiasm but get them to focus on how they can incorporate AI into individual process steps, rather than lose time rewriting a process over and over.

Failing Forward

Leadership expert John Maxwell helped make the phrase "Failing Forward" popular when his book *Failing Forward: Turning Mistakes into Stepping Stones for Success* sold about 30 million copies globally.[35] The book focuses on building a mindset of learning, persistence, and

---

[34] Dell'Acqua, F., McFowland III, E., Mollick, E., Lifshitz-Assaf, H., Kellogg, K. C., Rajendran, S., Krayer, L., Candelon, F., & Lakhani, K. R. (2023). Navigating the jagged technological frontier: Field experimental evidence of the effects of AI on knowledge worker productivity and quality. Harvard Business School. https://www.hbs.edu/faculty/Pages/item.aspx?num=64700

[35] Maxwell, J. C. (2000). *Failing forward: Turning mistakes into stepping stones for success.* Thomas Nelson.

progress, rather than focusing on failures as being setbacks. These tenets are especially powerful during times of significant change, as they can help people adapt more quickly and be more resilient.

In Chapter 3, we offered advice on how to build up organizational resilience, which included enabling and empowering improvisation. This is an extremely important extension of that thought, of the building of a culture of experimentation, trial and error, testing and checking, and generally not being afraid of failure. Even the creators of these tools only have broad ideas of their limitations and applications, so it will be the unique creative genius of your team that will matter most.

To foster this critical culture of failing forward in your organization, you might even need to step out of your own comfort zone. Shifting your company award ceremony to focus exclusively on risk-taking and noteworthy failures. Focus collaborative workshops on goals that can't be achieved. Create a charge code for "wild idea testing" and require 1 hour per week to be spent using it. Flood communication and strategic plans and office wall art with the goals, needs, and adjectives that put the concept of experimentation, innovation, and adaptation at the forefront of everyone's minds.

Also, don't forget to build a net strong enough to keep building resilience. Be ready to step in and stop those "just a few more days, and we'll get there" projects that look likely to drain resources and motivation indefinitely. Demand that post-mortem analyses happen after every failure to extract the most possible value. Identify and trace clear follow-up actions to update training, revise processes, or implement new strategies. Most of all, ensure supervisors at every level are openly encouraging, supportive, and willing to keep learning themselves.

Up the Slope of Enlightenment

You might recall from Chapter 3 that the Slope of Enlightenment is when the true value of tools becomes increasingly clear. The trough was a difficult time, but the lessons you learn from it and the cultural shift it enables will yield benefits for years to come. With any luck, your momentum from before the trough did a great job carrying you through, with only minor hiccups along the way.

Now, to move a broader audience out of the trough and up the slope, it's time to focus almost exclusively on targeting the late majority and laggards. Find them, help them, push them, and plead with them. Give them no place to run and no place to hide.

Consider setting new standards for promotions and annual raises tied directly to engagement and experimentation. Adjust organizational policies to highlight "how work is done here" and communicate these updates far and wide. Extend training and support offerings, and share these in the same emails that tout recent success stories.

You might also target them individually, asking supervisors for lists of slow adopters and including them in role swap initiatives. Bring them together for collaborative workshops facilitated by a patient and influential member of the change leadership team. Conduct a tailored survey only to these personnel and ask them to rank their biggest constraints and concerns, and then come up with other creative ways to address these based on what you learn.

This effort of finding and bringing along the slow adopters will be ongoing and likely indefinite, so don't stress if things don't work immediately. In the same way that the overall cultural objective is about adaptation and failing forward, driving toward incremental progress here is what's most important.

Soon, you should start observing sharing and teaching happening at multiple levels of the organization and new personnel leaning into the culture of experimentation and failing forward. The stories should be shifting from hyperbole to demonstrated successes. Some teams may already be experiencing some "Norming" as they find the limits of the current technology and naturally morph into centaurs and cyborgs in their daily work.

And yes, there's still work to do in the Norming and Performing phases, which we'll touch on in Chapter 6.

6: JUST KEEP SWIMMING

*"I'm convinced that about half of what separates successful
entrepreneurs from the non-successful ones is pure perseverance."*
- Steve Jobs

We said "touch on" Norming and Performing phases in Chapter 5,
and Tuckman would likely not be happy to hear we smashed in
Adjourning as well, all in the same chapter. We do have a few
thoughts and recommendations in these phases; however, this book
was intentionally not written about working through the challenges
of "steady state" technology adoption. As you've no doubt gotten
tired of hearing, our expectation is that this Fourth Industrial
Revolution will keep all of the most successful organizations very
much on their toes. You will be dancing from one tactic to the next
as tools and talent continually mature and evolve.

Perhaps the academics are already studying these trends and
working on a more applicable framework. We'll keep our fingers
crossed and ears to the ground for you.

The point is that if you were expecting to look out across the
great Plateau of Productivity and breathe a deep sigh of relief, you're
likely to find a massive "Rinse and Repeat" billboard blocking your
view instead. As teams start finding ways to adapt (Norming), getting
more comfortable with the change (Performing), and feeling
satisfied with achieving some of their goals (Adjourning), they will

already likely be facing the next challenge. A major update to shift their center of gravity again.

As a result, this chapter touches only briefly on three concepts to keep your organization nimble and prepared: monitoring progress, communicating success, and building up a strong feedback loop for continuous improvement. And don't peek yet, but Chapter 7 is where we speculate (perhaps wildly) on what steady state might look like for continuously learning organizations in the relatively near future.

Lightly Monitor Progress

In their May 2024 survey on AI deployment, Gartner found that "The primary obstacle to AI adoption ... is the difficulty in estimating and demonstrating the value of AI projects."[36] From our perspective, this may be one of those cases where the wrong questions were being asked. Yes, it's tough quantifying the value of today's generative AI, but that's because (1) the hype cycle promised more tangible and immediate results, and (2) we're seeing an entirely new kind of value being produced in modern knowledge-based work environments.

Things like "clearer, friendlier communications," "more focus on priorities," and "more effective brainstorming" have to be measured in broad, long-term impacts. Better indicators might be employee satisfaction and retention rates, more successful product launches, better returns on marketing campaigns, faster turnaround on continuous improvement efforts, fewer legal issues with contractual documents, less customer turnover, and fewer language barrier issues in foreign countries. To measure the true impact for your organization, you need to assess historical progress toward strategic objectives against post-AI progress. A daunting challenge if you don't have a good baseline of metrics to begin with.

---

[36] https://www.gartner.com/en/newsroom/press-releases/2024-05-07-gartner-survey-finds-generative-ai-is-now-the-most-frequently-deployed-ai-solution-in-organizations

If you're unable to monitor sweeping impacts effectively, or you don't really care as long the company keeps gaining ground, our recommendation is to track projects just enough to keep them moving. As Dr. Peter Drucker is famously credited with saying, "If you can't measure it, you can't manage it." We also like Mr. Tom Peter's "What gets measured gets done" from his 1982 book, In Search of Excellence.[37]

In monitoring the progress (Norming, Performing) and impacts (Adjourning) of generative AI projects, particularly if done for the sake of motivating progress (rather than demonstrating some specific outcome), here are a few recommendations to consider:

♦ Align with Long-Term Objectives – Create a clear connection between projects and strategic or other long-term objectives (growth, customer satisfaction, quality, efficiency, retention, etc.). This helps ensure projects are being picked and prioritized that lead toward desired end states and drives project owners to keep focus on those outcomes.

♦ Use Qualitative Prioritization Measures – To help prioritize projects based on impacts, use "high/medium/low" indicators to qualify the feasibility, level of effort, and potential impact of a project. Remember, you can always come back later and use generative AI to assess the values and make them quantitative for ROI calculation purposes.

♦ Conduct Periodic Surveys – Use broad-sweeping anonymous surveys (perhaps from a trusted third party) to specifically ask about engagement with generative AI. What is their level of engagement? Preferred method of use? Preferred tool? And what is their current job satisfaction? Anxiety level? Biggest area of concern? These can help you build valuable trends over time.

♦ Make Tracking Easy – Encourage progress monitoring for even small, quick-turn projects by asking for the fewest inputs possible. Who owns the project, what was done before, what are

---

[37] Peters, T. J., & Waterman, R. H., Jr. (1982). *In search of excellence: Lessons from America's best-run companies.* Harper & Row.

you trying to accomplish, what happened as a result. The goal is to keep them picking up new projects and failing forward, not create a new pile of busy work.

Respond to Wins

In the last three chapters, we've talked over and over about sharing success stories – establishing a mechanism for it in Chapter 3, broadcasting early successes in Chapter 4, and using email blasts to engage the Late Majority in Chapter 5. And if you haven't heeded our advice during other stages, then it's time to think about it. In a culture of failing forward, clear successes may be both a rarity and extremely exciting.

In addition to just sharing the news, when something exciting does happen, you want to be ready to take full advantage of it so that the Adjourning amounts to more than a quick pat on the back. How you accomplish this really depends on the size of your organization, the impact of the success, and how effective your cultural transition has been to date. In most cases, putting the project leader in a position to share what they learned with others is key.

Below are a handful of ideas you might seriously consider implementing in response to a success.

- Allocate additional resources to broaden the scope of the project and tackle bigger or related challenges.
- Have the project leader facilitate an innovation workshop to resolve a similar challenge.
- Create a more detailed training video that explains the challenges, process steps, breakthroughs, and results in a way that could be repeated on other projects.
- Have the project leader be a guest speaker at lunch and learn type training events.
- Create a discussion thread (facilitated by the project leader) to answer questions and make suggestions to others working on similar projects.

♦ Promote the project leader into a more senior technical role in the company where they can be the official "process owner" and help guide/influence other similar experiments.

Whatever you do, the point is to watch for and lean into big wins when they happen and make the most out of the intellectual capital that led to them. You likely already know what has historically worked for your team, so start with those approaches, see what happens, and then expand or try new things as needed.

Strategic Re-Alignment

In Tuckman's model, the Adjourning phase is one of celebration and reflection, but it also includes a connection back to the Forming phase where "Norms" are revisited. In addition to making the most out of what you learn from individual projects, it's important to look broadly at your organization as a whole. What types of projects are yielding the most value or the least? What teams are having the most success with experimentation? What are they doing that's working better than others? How are the tools evolving, and how are we preparing for the change?

Pro Tip: Hold a collaborative workshop to compare objectives and priorities to project data and survey feedback. Have the team determine the most likely sources of success and failure and ask for recommendations in areas such as strategic objectives, oversight, training, and communication to maximize the benefits of AI tools.

Answering these questions is difficult, and it will take a concerted effort. We strongly recommend working with generative AI to create a more comprehensive list of questions than we offer here. However, once you start getting even a few answers, you'll also start seeing some of the changes that need to be made. Below are some of the types of changes you may want to consider – this list may also help you identify the types of questions you need answered before you can make the right change.

- Encourage shifts in experimentation away from high-failure, low-reward areas and toward high-success, high-reward areas.

- Present high-priority challenges to top-producing project managers and teams, regardless of their functional area of expertise (remember, generative AI helps bridge that gap).

- For areas of high success and high reward, assign research projects to identify, get demonstrations, and test out Tailored Technologies that offer another leap in capability.

- Shift survey questions to prioritize and capture more of the insight that's driving strategic changes and less of the insight that's not getting used.

- Revise recruiting, onboarding, and training programs to prioritize the backgrounds, education, and communication styles that have achieved the highest success levels.

- Formally shift work processes, team make-up, and leadership to best take advantage of innovations. If that's too difficult to do, consider the creation of centers of excellence for topical cross-team sharing and collaboration.

- As you see what's possible today and what might become possible in the future, take the time to update your strategic plans and corporate objectives.

7: INTO THE AFTERMATH

"The delta between 4 and 5 will be the same as between 4 and 3."
– Sam Altman, OpenAI CEO, March 2024

Thus far, we've given you a lot of great suggestions and advice on how to prepare your team for the future. But what are you preparing for? How much change can we really expect?

Consider for a moment the impact that e-commerce has had. In the early 2000s, e-commerce was on very few radars (1% of retail sales in 2000), and it's now a major and growing aspect of nearly every business (16% in early 2024).[38] To accomplish this, companies have invested heavily in IT infrastructure, dedicated online sales teams, customer service departments, advanced data algorithms for personalized recommendation systems, sophisticated and automated inventory management systems, and much more. Sales teams have global customer bases and contend with new international cultural and regulatory challenges. Significant shifts in retain landscapes occurred, too, such as the closure of Borders, Toys R Us, and Sears, while e-commerce-focused companies like Amazon, Alibaba, and Wayfair have experienced unprecedented growth.

No one can perfectly predict what changes your company will

---

[38] https://fred.stlouisfed.org/series/ECOMPCTSA

experience due to generative AI, but moving forward on educated guesses will certainly minimize the potential risk. To start answering the questions we posed above and to help position you to take advantage of these technologies as they evolve and emerge, this chapter takes a speculative look at what highly competitive organizations might look like in the future. We touch not only on overall operations but impacts on your organizational structure, your workforce, and the tools you may have accessible in the near future.

For your sake, we don't pull any punches here, even when the changes seemed almost too wild to write down. Our hope is seeing these potential changes will deepen your sense of urgency and inspire you to take some of the suggestions in the previous chapters even more seriously. So, brace yourself and enjoy.

The Death of Organizational Inertia

Organizational inertia is the bane of executives everywhere. It's the ever-present sludge of "how it's always been done" that resists innovation, slows down adaptation, and enables the persistence of inefficiencies. But the cultural ecosystem that enables it, from communication to team structures to supervision to resource allocation patterns, is going to be changing drastically. Put another way, how work gets accomplished in the future will be fundamentally different. Below are just a few examples of what will very likely be possible.

Full organizational alignment. Today's AI tools improve the clarity and relevance of communication, and they are already able to break down strategic plans into specific actions for every level of an organization. Future-proof companies will instantaneously shift to new priorities – resources allocated, projects initiated, tasks assigned and understood, and performance measures in place. At the moment one growth strategy shows signs of stagnation, a new one will be rolled out. This will be enabled by not only streamlined communication tools but also fully adaptable expertise, self-organizing teams, and an instant-response supply chain of companies adapting just as quickly.

Hyper-flexible, hyper-focused project execution. The future will see the rise of an interest-based culture where your team of fully adaptable experts jump into and out of projects that meet their "passion profile." AI will help solve the supervisory inefficiencies that exist today between deciding on actions needed, determining which resources are available, getting them up to speed on what's needed, keeping tabs on progress, and validating that their work supports the overall goal. Once these things can be accomplished in real-time, project managers and supervisors will shift focus to improving the job satisfaction and quality of life of their teams, helping them find and refine their work-related passions.

AI Process Owners. Today's generative AI tools are already more capable of process design than most humans, so it's no stretch to say that the documentation and management of processes will be changing dramatically in the near future. In fact, Frontier Academy began work on a process improvement guide, but the book was becoming outdated as fast as it was being written. We predict that Standard Operating Procedures, Work Instructions, Manuals, and Guidelines will all soon be historical artifacts, relics of a time when they were needed to ensure consistent quality of results. The next step in the process will be recommended based on a clear understanding of the overall goals, the project or team objectives, and what the individual contributor needs to do next to help.

Rapid testing and integration of new tools. A theme of this book is building your culture of experimentation, and with hundreds of new tools being released each month (as noted in Chapter 2), there will be many that make a big difference in your bottom line. Not only will more tools exist, but their capabilities will continue to improve exponentially along with the AI models they're based on. Companies that remain competitive will be in a constant state of testing what will have an impact and rolling it out.[39] Either that, or you'll find yourself quickly giving up market share while you wait for the perfect tool to come out.

---

[39] https://mitsloan.mit.edu/ideas-made-to-matter/how-ai-helps-acquired-businesses-grow

A supply chain that predicts your needs before you ask. The extreme progress being made in robots and manufacturing automation is not slowing down. The robotics industry generated ~$38B in 2023 and is projected to exceed $43B by 2027.[40] In fact, literally while writing this paragraph, we learned that Figure's "01" robot is starting to help BMW build cars in South Carolina.[41] Soon, manufacturers will be monitoring market trends along with you, shifting outputs in rapid response and even in anticipation of your next request. And when they can't keep up, new companies (who saw the need and took action) will appear immediately to take their place.

Shapeshifters Wanted

The power of today's generative AI lies in its peculiar ability to act as an almost expert in almost every subject, and as this power grows, we will see an increasingly wide divide between the individuals who embrace it and those who do not. The divide will take the form of not only productivity but also communication proficiency, problem-solving capability, intellectual curiosity, and the ability to successfully take on responsibilities far outside of their experience or education.

As you take advantage of this divide and increasingly employ AI-proficient team members, your workforce will become more focused on what matters and have more depth of expertise and resilience to change. As mentioned above, your new resource economy will be driven by passion and interest, not just capability and availability. To enable this change, premiums will be placed on Meta skillsets (explained below), and there will be a thriving freelance marketplace readily available to meet surges in needs as you grow. There's a lot to unpack here, but the paragraphs below attempt to at least briefly explain how all of this will be possible.

---

[40] https://explodingtopics.com/blog/robotics-industry-stats

[41] https://interestingengineering.com/innovation/us-figure-humanoid-start-operations-at-bmw-plant

Increased focus on what matters. Already, we are seeing major changes in the employment landscape for the most automate-able skills. According to the Institute for Public Policy Research (IPPR)'s March 2024 study on the impacts of generative AI on work in the UK, the "here and now AI" is already capable of automating cognitive jobs with repeatable, consistent tasks (administrative assistants, customer service, marketing, translators, etc.).[42] The flip side of this scary coin is that jobs will become increasingly interesting. As AI helps us pick up more tedious tasks and begins to help with the bigger challenges, super-charged humans will have more and more impact on their jobs, dramatically improving employees' sense of self-worth, value, and overall job satisfaction.

Depth of expertise and resilience to change. The North Carolina Department of Commerce reported in February 2024 that "the most significant benefits in productivity were observed among less-experienced, lower-skilled workers."[43] This trend is enabling historically low-level performers to surge ahead and approach the productivity of higher-level performers. The reason is that generative AI offers every user access to a wealth of understanding, akin to a subject matter expert in every field waiting beside you, eagerly ready to answer your questions. As AI models continue to improve, so will the level of expertise and ability to anticipate user's questions, eventually giving every motivated employee the ability to perform well in any knowledge-based role in a company.

A resource economy driven by passion and interest. Once functional assignments are no longer limited by our experience or education, how will resource allocations occur? This is where our "passion profile" and "interest-based culture" phrases stemmed from. If a knowledge worker can be moved into any function, any good supervisor knows to put them where they'll be the most passionate and interested. This will require individuals to better

---

[42] https://ippr-org.files.svdcdn.com/production/Downloads/Transformed_by_AI_March24_2024-03-27-121003_kxis.pdf

[43] https://www.commerce.nc.gov/news/the-lead-feed/generative-ai-and-future-work

understand, keep up with, and be able to communicate their own evolving areas of interest. We may see self-organizing, high-powered micro teams form, where a handful of motivated individuals take on one major corporate challenge after another. It will all depend on the strategic goals of the company and the degree of flexibility the leadership and culture empower.

Meta skillsets as the new job currency. As generative AI begins to serve as the great translator and teacher of cognitive workers, education and experience will become increasingly useless indicators of competency. In fact, this is already the case. According to Hiring Lab (funded by Indeed), "mentions of college degrees have fallen since 2019 in 87% of occupational groups."[44] As AI touches more and more job functions, the new premium skills will increasingly be those needed to work with AI and adapt to new roles, such as flexibility, communication, and emotional intelligence. If you aren't already adding the "ability to communicate effectively with AI tools" to your job requisitions, it may be time to think about it.

Rest assured, your next generation of employees is already unconsciously preparing for this. As an example, Character.ai (a primarily entertainment platform for chatting with AI-driven personalities and experts) data shows that 60% of their 200+M monthly users are between 18 and 24 years old. [45] And if your current workforce doesn't develop these skills, we anticipate that the need for them across all industries will produce a wave of freelancers ready to jump in as consultants for a day or a year, provided the project matches a passion of theirs.

Corporate DNA Resequenced

Structurally speaking, we anticipate corporate organizations will look much different as they shift to take advantage of these new capabilities. Faster decision-making and response times will drive

---

[44] https://www.hiringlab.org/2024/02/27/educational-requirements-job-postings/

[45] https://whatsthebigdata.com/character-ai-statistics/

shorter strategic planning cycles. AI support to performance management and resource allocation will reduce the need for middle management and enable Human Resources teams to handle some of the big new changes. Sales teams will become smaller and dramatically more impactful. And a host of new teams and functions will appear to keep this new engine running smoothly. This section offers our conjecture on what some of these teams will look like.

Market Response Team. A small team of executives will be responsible for market analysis, quick-turn strategic planning (think quarterly or even monthly), and full-scale implementation planning for the entire company. This includes project identification, planning, resource allocation – the whole gamut. Generative AI tools are already mostly capable of supporting this need today, and tailored technologies will make this infinitely more accessible going forward.

Passion Management Team. What was historically managed by layers of managers and supervisors will be a handful of middle managers able to take care of hundreds of personnel. They will be the ones validating passion profiles and job satisfaction and just making sure people have what they need to be successful. They will be the faces in front of powerful generative AI tools being used to check in with folks and monitor their emotional well-being. If you're surprised to hear that AI might help with this, know that early versions of AI have already proven themselves to have much higher empathy than humans. In fact, in a 2023 study comparing physician to chatbot responses, they found that AI had a "9.8 times higher prevalence of empathetic or very empathetic responses."[46]

Talent Pipeline Team. We think it likely that your Human Resources team will need to remain largely intact, but rebranded and recalibrated for a host of new duties as many of today's functions become more automated. AI will be able to handle the impersonal stuff (combing through applications, processing paperwork, checking onboarding status, monitoring performance, and more). But with these functions off the table, your Talent Pipeline Team will become much more focused on helping employees increase their

---

[46] https://pubmed.ncbi.nlm.nih.gov/37115527/

Meta skills, monitoring market trends in labor, assessing the value of full-time hires against gig workers, ensuring ethical and legal use of AI tools, and shifting passion profile needs in response to strategic planning updates.

Client Relationship Team. Where today's sales teams spend the majority of their time managing client data, trying to chase new leads, and planning for big calls, that daily grind will be mostly automated. Sales teams of the future will be able to handle significantly higher volume, working with vetted leads and being "the face" of the hyper-personalized offering that the prospect has already received. AI will help monitor prospect body language in real-time, reducing communication breakdowns and ensuring un-asked questions get answered. In many cases, sales may even be conducted AI-to-AI in advance of human-to-human meetings, drastically improving closure rates. But there will likely always be the need for in-person meetings, simply because human customers tend to prefer talking to other humans. The 2024 Restaurant Technology Landscape Report validated this sentiment, noting that only 33% of adults are currently in favor of talking to AI to place orders at a restaurant.[47]

Technology Deployment Team. You will likely need dedicated resources simply to keep up with available tools and technology to keep your Market Response Team executives aware of the potential changes and efficiencies. A ZoomInfo study of 2021 to 2023 has already shown a dramatic rise in senior leadership (240%) and technical jobs (5,000%) with AI titles in them as companies seek to take advantage of these tools.[48] This critical team will be responsible for finding, testing, and implementing technology updates as quickly as possible, always trying desperately to keep up with competitors (and customers) who are doing exactly the same thing.

Data Quality and Reliability Team. One increasingly important area of focus will be data management. To be maximally

---

[47] https://go.restaurant.org/rs/078-ZLA-461/images/
NatRestAssoc_TechLandscapeReport_2024.pdf
[48] https://pipeline.zoominfo.com/sales/the-ai-boom-in-senior-leadership-jobs

useful, generative AI tools need standards and insight that will make them more tailored to your niche area of business. In their 2024 *AI Business Predictions* report, PWC found that "44% of business leaders said that their companies are planning to implement data modernization efforts in 2024 to take better advantage of GenAI."[49] How you feed your AI systems beast will require a very intentional, systematic, and systemic gathering, storing, and quality control data management strategy that allows for rapid shifts in demand, scalability, and integrated access with AI tools.

Tomorrow's Tools

While our computers and smartphones have certainly gotten faster and more capable over the years, the vast majority of today's workforce is barely scratching the surface of productivity technologies. As you think about future-proofing your workforce, we recommend looking at not only the near-term impacts of generative AI but also the long-term potential for an infusion of tools that could be game-changing for your company and industry. We picked just a handful of examples to illustrate our point, but there are, of course, many, many more.

We apologize in advance for all the citations embedded in this section, but we thought it was important to show how "here" and "nearly here" these technologies really are.

Agentic AI Tools. As generative AI becomes more capable and more integrated with our daily work life, we will likely see it begin to run out ahead of us.[50-51-52] Meaning, it will anticipate and begin working on the next step without being asked. The implications are massive, not just for individual productivity but marketing,

[49] https://www.pwc.com/us/en/tech-effect/ai-analytics/ai-predictions.html

[50] https://blogs.microsoft.com/blog/2024/05/20/introducing-copilot-pcs/

[51] https://www.businesswire.com/news/home/20240711308554/en/Peak-Boosts-Business-Productivity-with-General-Release-of-Agentic-AI-Assistant-CoDriver

[52] https://wallstreetpit.com/119304-beyond-human-limits-gpt-4os-reading-speed-rewrites-ai-possibilities/

production, customer service, and more. Think about having real-time changes being made based directly on market indicators.

Extended Reality Systems. Virtual Reality, Augmented Reality, and Holographic displays promise immersive remote collaboration and improved visualization of complex concepts.[53][54][55][56] These tools are increasingly present and adding value to our physical space in measurement, presentations, training, and much more. They may very well be the solution to the ongoing conversation about remote versus in-person work requirements.

Emotion Interpretation Systems. Emotional AI and Sentiment Analysis Systems will change the face of communication, reducing errors in understanding and improving human-to-human empathy.[57][58][59] In areas like team dynamics, sales engagements, and customer service, we will see better collaboration, more productive conversations, higher satisfaction, and the list goes on.

Brain-Control Interfaces (BCI). Scientists are developing neuro-enhancing chips and non-invasive nanobots to enable connectivity to external computing capabilities.[60][61][62][63] By interfacing

[53] https://program-ace.com/blog/the-impact-of-virtual-reality-on-the-business-world/

[54] https://www.business.com/articles/best-augmented-reality-uses/

[55] https://www.forbes.com/sites/moorinsights/2023/06/19/the-rise-of-augmented-reality-in-the-modern-workplace/

[56] https://engineering.princeton.edu/news/2024/04/22/holographic-displays-offer-glimpse-immersive-future

[57] https://www.forbes.com/sites/josipamajic/2024/01/30/ai-empathy-emotional-ai-is-redefining-interactions-in-the-digital-age/

[58] https://www.nojitter.com/ai-speech-technologies/promise-and-perils-sentiment-analysis

[59] https://www.cxtoday.com/contact-centre/the-development-of-sentiment-analysis-how-ai-is-shaping-modern-contact-centers/

[60] https://academic.oup.com/pnasnexus/article/3/2/pgae076/7609232

[61] https://thedebrief.org/the-fda-has-approved-neuralinks-brain-chip-implant-technology-for-a-second-patient/

[62] https://www.asiafinancial.com/china-unveils-brain-chip-similar-to-elon-musks-neuralink

[63] https://onlinelibrary.wiley.com/doi/full/10.1002/smsc.202300211

your neural network, these technologies offer not only thought-to-command capabilities but also increased focus, creativity, and the rapid ability to understand complex concepts. Think about the impacts to your business of eliminating language barriers and learning curves.

And there's a lot more out there. In power, there may be radioisotope-powered "batteries" that enable almost continuous run time for electronic devices.[64] In logistics, distances may soon be tackled by increasingly viable, AI-powered autonomous vehicles.[65] In health, gene editing techniques may be used to resolve genetic defects.[66] So keep your ear (or your research bots) to the ground and stay ready to consider how new advancements can help impact your company's trajectory.

Tomorrow's Humans

In 2023, at a Berkeley Haas event, NVIDIA CEO Jensen Huang stated, "For the very first time in our history, in human history, biology has the opportunity to be engineering, not science. When something becomes engineering, not science, it becomes less sporadic and exponentially improving."[67] In his 2024 book, *The Singularity is Nearer*, renowned futurist Ray Kurzweil refers to "people with cloud-connected neocortices" and "radically new possibilities for how the brain itself processes experiences" (like clearly understanding an actor's nonverbal thoughts and emotions while watching a movie and co-creation of new ideas with AI superintelligence). He defines the Singularity as the point in time where human capabilities will exceed our current understanding of what's possible.

---

[64] https://www.techradar.com/phones/a-tiny-radioactive-battery-could-keep-your-phone-running-for-50-years

[65] https://blogs.nvidia.com/blog/waabi-autonomous-trucking/

[66] https://www.news-medical.net/news/20240711/New-gene-editing-technique-offers-hope-for-millions.aspx

[67] https://www.youtube.com/live/9hzVdV63scU?si=2tyQz5XF8HsZxv-e&t=2956

If Ray Kurzweil can't predict what this means, we're certainly not going to try. We can only hope that continually preparing your workplace culture will help set the stage for handling augmented humans as part of your workforce. They might redefine performance metrics and have instantaneous learning curves. They might be genetically engineered to live forever. There may be a new divide in the workforce between augmented and traditional humans that makes today's generational differences seem insignificant. And there may also be a host of new legal, regulatory, and ethical issues to contend with. Our advice here is simply to be aware that these types of advancements are being researched and tested, and they will eventually have profound implications on what humans are ultimately capable of.

Okay, So Now What?

First of all, take a deep breath. Remember, the purpose of this chapter was to give you a burst of urgency-laced adrenaline. After a couple of nights' sleep, these changes will once again feel like a distant dream of a future.

However, the next time you read this book, and we do recommend a second read, you will hopefully read it with a renewed passion. With any luck, that second read is when you'll really start writing down some action items to help get your team started on this important journey.

8: A BRIGHT FUTURE

"These are the most exciting and momentous years in all of history."
– Ray Kurzweil, *The Singularity is Nearer*, 2024

As short as this book is, it's a lot. A lot of change. A lot to do. And a lot of long-term strategies to consider. It's frightening and exciting at the same time.

Recall Sarah's story of a not-so-distant future. It's an imagined future (at this point) and will materialize differently in reality, but we strongly encourage you to start imagining your own version. Already today, companies of all sizes are leveraging AI to reduce overhead costs, improve productivity, and find creative ways to expand. Soon, these tools will be as essential as the computer and the internet, helping new employees leap forward in their careers and helping seasoned experts shape the future more than ever before.

How do you see the technology evolving? What steps can you take today that will best position your company for these changes? How can you prepare your company to react when the technology inevitably shifts direction again?

Our minimum recommendation is to start holding serious internal conversations about what's next for your company. Consider creating a simple "AI Strategy" and get buy-in to start working toward it. Help other executives learn to speak intelligently about the tools and technology. Maybe even start finding some early adopters in your organization to help.

Whatever you do, we hope this book serves to give you a renewed sense of urgency – to avoid being swept aside by these changes and be ready to leverage them for growth in your markets.

THANK YOU

As a small token of our appreciation for your purchase of this book (and for that great Amazon review you almost forgot about), below are a couple links to help you get started.

Future-Proof Readiness Assessment

The link below will take you to a Custom GPT designed to help you assess your organizational readiness for generative AI integration. Answer a few questions, and it creates a simple action plan designed to motivate you and your team toward taking additional steps.

https://chatgpt.com/g/g-Mflh5AC6V-future-proof-readiness-assessment

Strategic Plan Review

Click below for a Custom GPT that reviews strategic plans and offers simple tweaks to better align your corporate objectives with the future-proofing we believe is needed for continued growth. Its suggestions are meant to get you thinking and get you started.

https://chatgpt.com/g/g-677e7e15c0048191b828ce1a2d6ae474-future-proof-strategic-plan

ABOUT THE AUTHORS

Raymond Vogel is a seasoned professional with over 20 years of experience as a business leader, sales expert, writer, and engineer. Throughout his career, Raymond has been at the forefront of innovation and continuous improvement, consistently challenging the status quo to drive efficiency and growth. His entrepreneurial spirit led him to build a successful publishing company, while his technical background enabled him to develop innovative automation solutions across various industries. As a thought leader in productivity enhancement and process optimization, Raymond brings a unique perspective on leveraging AI tools to revolutionize workplace dynamics. His passion for empowering teams and individuals makes him an ideal guide for professionals navigating the rapidly evolving landscape of AI in business.

Dina Alkhateeb is a dynamic leader in corporate management and communications with a proven track record of driving high-performance teams to success. Her career has been marked by a relentless pursuit of excellence and a productivity mindset that has consistently delivered results in fast-paced, high-stakes environments. As an early adopter of emerging technologies, Dina has always been at the cutting edge of integrating new tools and methodologies to enhance workflow efficiency. Her experience spans from hands-on technical writing to executive leadership roles, giving her a comprehensive understanding of how AI can transform various aspects of business operations. Dina's expertise in developing and implementing innovative solutions makes her an invaluable resource for professionals seeking to harness the power of AI in their work.